Bibliographic information published by the German National Library:

The German National Library lists this publication in the National Bibliography; detailed bibliographic data are available on the Internet at http://dnb.dnb.de .

Imprint:

Print and binding: Books on Demand GmbH, Norderstedt Germany
ISBN: 9783668270817

This book at GRIN:

http://www.grin.com/en/e-book/335318/hydrochemistry-of-water-from-bauchi-state-nigeria

Josiah Jabbo

Hydrochemistry of water from Bauchi state, Nigeria

GRIN Publishing

HYDROCHEMISTRY OF WATER FROM SOME VILLAGES IN ALKALERI LOCAL GOVERNMENT AREA, BAUCHI STATE, NIGERIA

T. P. Bata[1], N.K. Samaila[1], J.N.Jabbo[2] and O.A. Solomon[3]

1. Department Geology, Faculty of Science and Technology, Abubakar Tafawa Balewa University Bauchi. Nigeria
2. Department of Geological Technology, School of Science and Technology, Abubakar Tatari Ali Polytechnic Bauchi. Nigeria
3. Department of Geology, Faculty of Natural Science, University of Jos. Nigeria

Abstract

Groundwater samples were collected from some villages in Alkaleri Local Government Area of Bauchi State. The samples were analyzed at the Geology Programme laboratory, Abubakar Tafawa Balewa University Bauchi. Concentration of iron, lead, manganese, chromium, calcium, copper, zinc and nickel were determined using the atomic absorption spectrometry (AAS). Results of the analyses show low values for iron, lead, manganese, calcium and copper in some locations (below the World Health Organizations (WHO) maximum permissible limit for drinking water). Concentrations of nickel and chromium in some locations are high (above the WHO maximum permissible limit for drinking water). Adequate concentrations of some of these trace elements in drinking water help in maintaining metabolism of the human body, however, higher or lower concentrations could have adverse effects on humans.

Inhalt

1.0 Introduction:

The area of study covers the villages of Gar, Panti, Badara, Vulum, Yashi and MainaMaji in Alkaleri Local Government of Bauchi State, which fall under the Gongola sub-basin of the Upper Benue Trough (see figure 1). This Gongola sub-basin is located NE of the Upper Benue Trough, and is a NNE-SSW trending rift about 800km long and 150km wide containing up to 6,000M of Cretaceous – Tertiary sedimentary rocks. The section pre-dating a mid-Santonian compressional phase has been folded, faulted and locally uplifted, resulting in over 1000 anticlines and synclines (Benkhelil, 1989). According to the works of Carter et al., (1963), the Bima sandstone is the earliest sedimentary rock deposited in the basin. The successive cretaceous rocks overlying the Bima in the Gongola sub-basin are divided in to the following formations which range in age from Cenomanian to Maastrichtian. Figure 2 below shows the stratigraphic descriptuion of the Benue Trough.

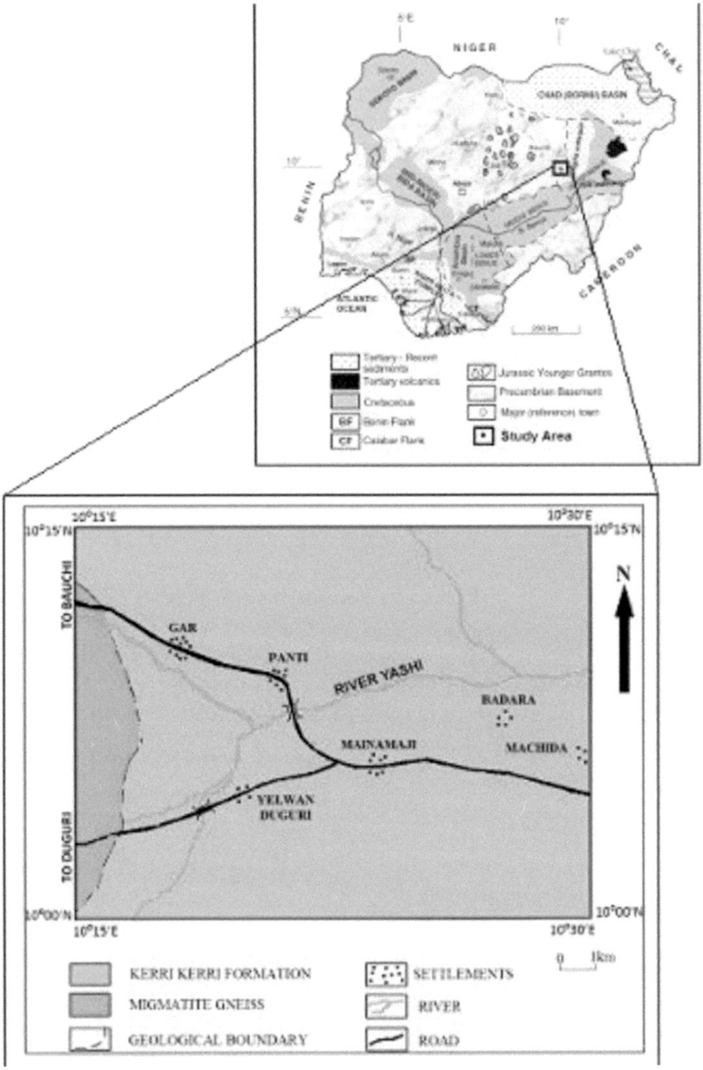

Figure 1. Geologic Map of Nigeria (minimized) showing the study area. (Modified from Obaje 2009

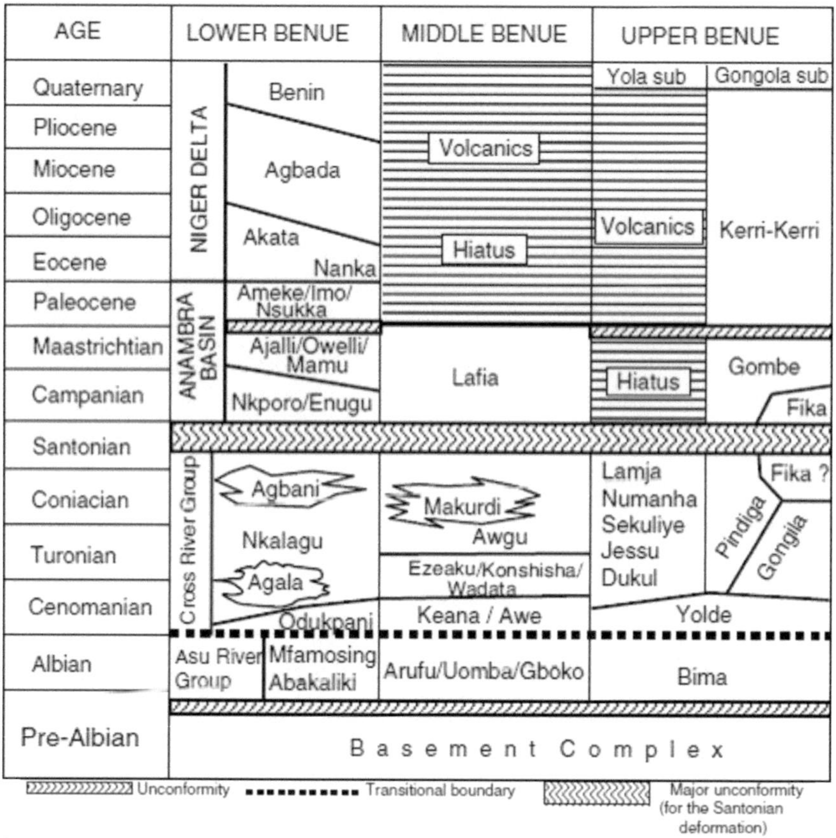

Figure 2. Stratigraphic Description of the Benue Trough.Adapted from Obaje (2000).

Groundwater in the study area occurs within the Paleocene Kerri-Kerri Formation. This is composed of poorly sorted sandstone with interstitial clays (Bata et al, 2009).

2.0 Method of Study.

A total of seventeen (17) ground water sample were collected from the study area. These samples were analyzed at the Geology Programme laboratory, AbubakarTafawaBalewa University Bauchi. Concentration of iron, lead, manganese, chromium, calcium, copper, zinc and nickel were determined using the atomic absorption spectrometry (AAS). The sample were digested first , by measuring 50ml of the water samples into the sample vapour, is illuminated by a light source calibrated beaker. Add 5ml of nitride acid (HNO3) and Hydrochloric acid (HCL) in the ratio of 3:2 (3ml of HNO3 + 2ML OF HCL). Heat the mixture on a hot plate until the volume is reduce to 10ml through evaporation and make it up to the mark in 50ml in volumetric flask with distilled water. Read the absorbance from Atomic Absorption Spectrophotometer (AAS), which will indicate the corresponding percentages for each traced element.

The Atomic Absorption Spectrophotometer (AAS) works on the principle that atoms in a flame at ground state (to be analyzed) absorb the same wavelength of light that will emit if exited. The sample is brought into solution. This solution is aspirated (dispersed or evaporated) by means of a flame (mixture of air and acetylene).which gives a temperature about 2000^{0}c. This high temperature reduces the droplet into atomic form I.e. the element in solution are now in vapour and therefore in natural or unbound state. This condition is to be on ground state". If higher temperature is needed to aspirate the solution, the flame would be made of mixture of nitrous oxide, and acetylene and it will give a temperature of up to 8000^{0}c. The sample in vapour is illuminated by a light source which is the hallow cathode lamp (HCL) is the as the material to be analyzed e.g. for the analysis of (Zn, Pb, Ca, Cr, Mn, Ni, Cu, Fe,) HCL was used so that the atom in vapour absorb radiation from the HCL. This radiation is isolated by means of monochromoter. The radiation from the HCL is reduced because of the absorption of vapour and a detector measures it. The amount of absorption (intensity) is proportion to the concentration of the element in the sample (vapour), Hence, by measuring the decrease in energy received by the detector, the amount of the element in the sample can be established (Skoog, et al. 1992).

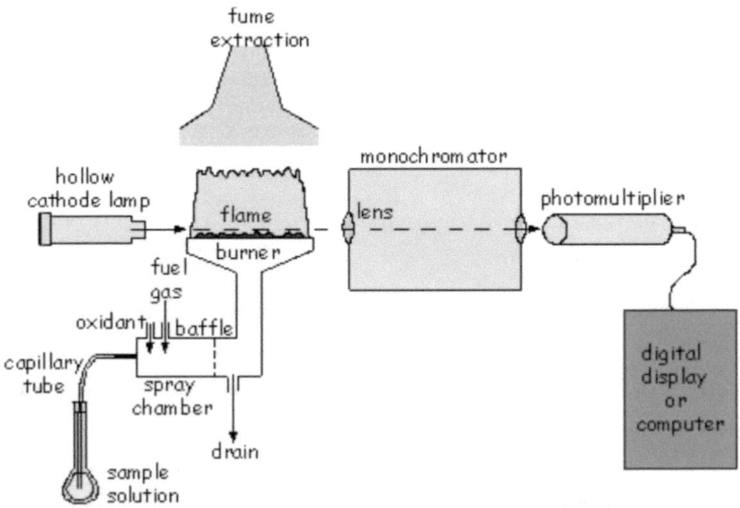

Figure 3 Schematic diagram of atomic absorption spectrophotometer (after Skoog et al, 2009)

3.0 Results and Discussions:

As mentioned earlier, seventeen (17) ground water samples were taken from four villages in the study area. The results are presented in figures 4 to 11.

Iron (Fe)

Iron is an essential part of hemoglobin; the red coloring agent of the blood that transports Oxygen through our bodies. Deficiency of Fe in humans can result in anemic conditions. A man needs an average daily intake of 7mg of Fe and a woman 11mg (Lenntech, 2009). Figure 4 is a chart showing the concentration of iron in drinking water of the study area. This chart shows that iron is deficient in drinking water of the study area. . Although iron is both plentiful and obtainable from a wide variety of foods, iron deficiency is still the most common problem in humans (Edmunds 1996). Children having iron deficiency may have difficulties in learning. People living in the iron is deficient areas should be encouraged to eat foods rich in iron such bone marrow, spleen, kidney and liver.

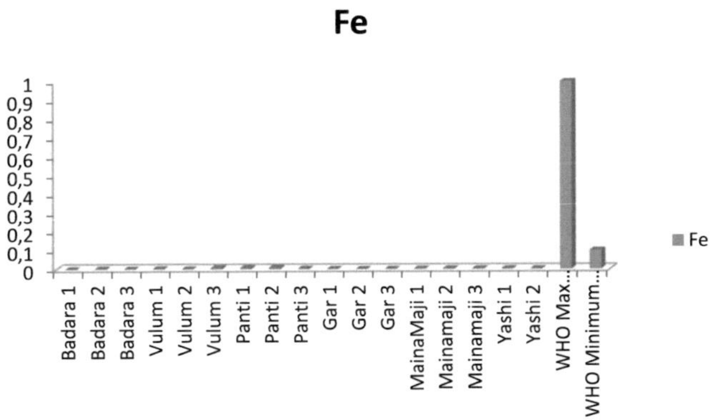

Figure 4. Chart showing the concentration of Iron in drinking water of the study area.

Lead (Pb)

Figure 5 shows the concentration of lead in drinking water of the study area. This chart shows that the concentration of lead is below the WHO minimum permissible limit; however this deficiency of lead has no negative effect on the people in the study area.

Copper (Cu)

Copper in many forms, is the third most common mineral in the body. However excess copper in the body can cause anemia, liver and kidney damages (Skipton et al 1988) amongst many other diseases. People in the study need not to worry about excess copper, because figure 6 shows that copper is deficient in drinking water of the study area. Therefore people of the study area should be encouraged to eat more of avocados, molasses, nuts, fruits and legumes to supplement for the copper deficiency

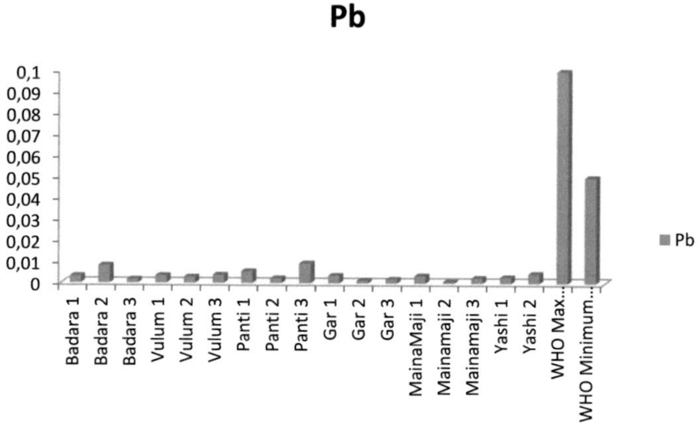

Figure 5. Chart showing the concentration of Lead in drinking water of the study area.

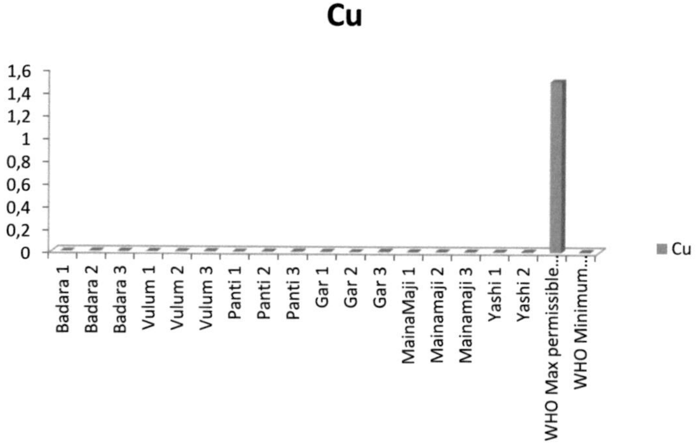

Figure 6. Chart showing the concentration of Copper in drinking water of the study area.

Manganese (Mn)

Manganese is seldom found alone in a groundwater. It is frequently found in iron-bearing waters but is more rare than iron. Chemically it can be considered a close relative of iron since it occurs in much the same forms as iron. When manganese is present in water, it is every bit as annoying as iron, perhaps even more (free drinking water.com 2010).

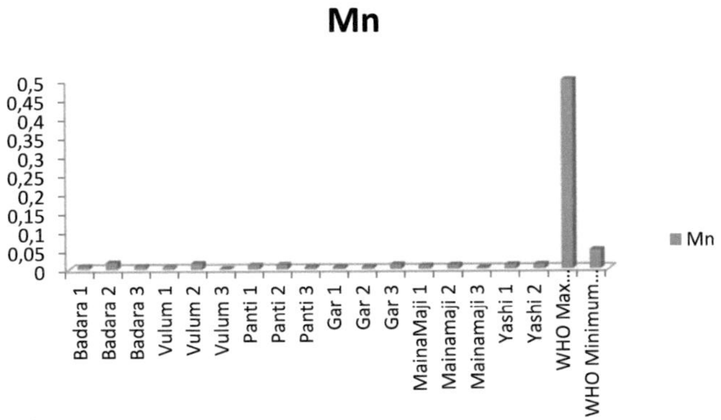

Figure 7. Chart showing the concentration of Manganese in drinking water of the study area.

Figure 7 is a chart showing the concentration of manganese in drinking water of the study area. This shows that manganese is deficient in the whole of the study area. Lenntech (2009) explained that deficiency of manganese (as in the study area) can cause adverse health effects such as fatness, blood clotting, skin problems, lowered cholesterol levels, skeleton disorders, birth defects and change of hair color.

Chromium (Cr)

Chromium is an essential nutrient for the human body, and shortages may cause heart conditions, disruption of metabolisms and diabetes. Too much intake of chromium on the other hand can cause various health effects like respiratory problems, stomach upsets and ulcers, weakened immune systems, kidney and liver damages, alteration of genetic materials, and lung cancer (Lenntech, 2009).

Figure 8 is a chart that shows the concentration of chromium in the study area. People living around Badara and Mainamaji are exposed to high concentration of chromium in their drinking water. This can cause various health effects like respiratory problems, stomach upsets and ulcers, weakened immune systems, kidney and liver damages, alteration of genetic materials, and lung cancer. People living around Gar village have chromium being deficient in their drinking water. As stated earlier, this may cause heart conditions, disruption of metabolisms and diabetes.

Cr

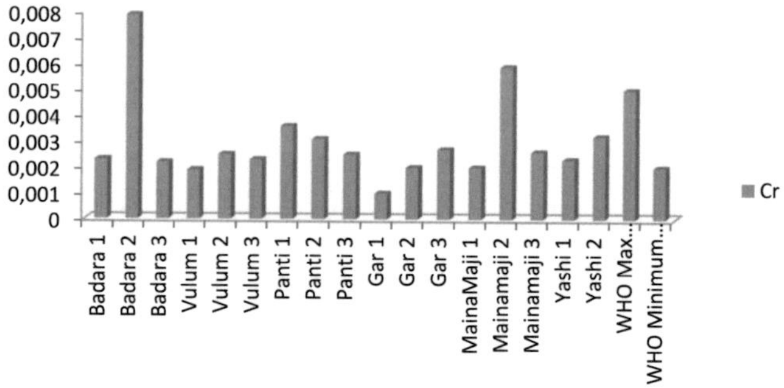

Figure 8. Chart showing the concentration of Chromium in drinking water of the study area.

Zinc (Zn)

Zinc plays a vital role in the normal functions of our body. It is required for growth, metabolism and hemoglobin build. Zinc is an essential trace element for human health. The daily requirement for an adult human is 15-22mg per day (WHO, 1997). When people absorb too little Zinc they can experience a loss of appetite, decreased sense of taste and smell, slow wound healing and skin sores. Zn shortage can also cause birth defects (Lenntech, 2009).

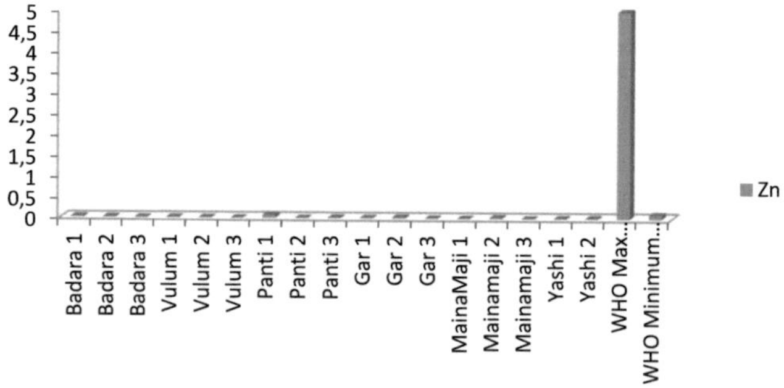

Figure 9. Chart showing the concentration of Zinc in drinking water of the study area.

Figure 9 above shows that zinc is deficient throughout the study area, Studies suggest that such low levels of zinc (Lantech, 2009) can slow healing and cause serious digestive problems that may result into chronic diarrhea. Many also believe that insufficient levels of the mineral can lead to longer recovery times from colds. Nursing mothers that have zinc deficiency may not give their babies the nutritional benefits of zinc through milk. People living in the study area should therefore be encouraged to eat nuts, grains, and beans. Milk, yogurt, and cheese are also excellent sources of zinc.

Calcium (Ca)

The human body needs calcium to develop strong teeth and bones. With that, calcium can combat osteoporosis and other bone disorder. It also helps in regulating nerve transmission, blood coagulation, and muscle contraction. (WHO, 1997). Figure 10 below shows that calcium is deficient throughout the study area.this means that peole living in the study area are exposed to health conditions such as *'Osteoporosis'* This is a disease in which the bones become extremely porous, easy to fracture, and heal slowly, occurring especially in women following menopause (Lenntech, 2009). Another common health effect of Calcium shortage is poor development of bones and teeth in both humans and animals.

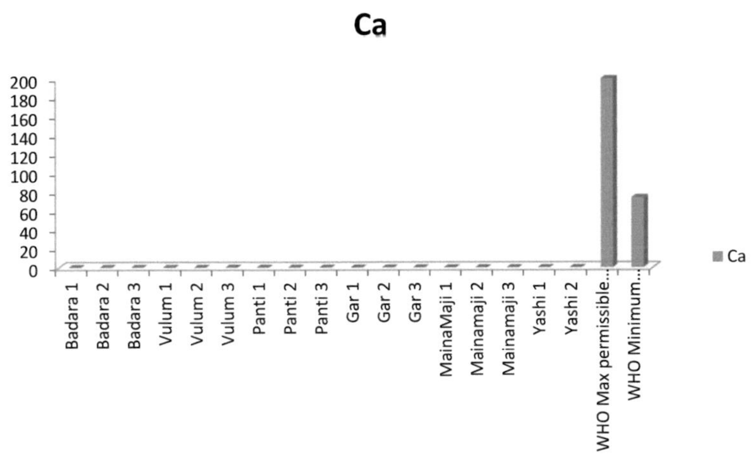

Figure 10. Chart showing the concentration of Calcium in drinking water of the study area.

Nickel (Ni)

Small amounts of Nickel are needed by the human body to produce red blood cells, however, in excessive amounts, can become mildly toxic. Short-term overexposure to nickel is not known to cause any health problems, but long-term exposure can cause decreased body weight, heart and liver damage, and skin irritation (Lenntech, 2009). Figure 11 shows the concentration of Nickel in the study area as compared to WHO minimum and maximum permissible limits. This shows that people living around Badara, Vulum, Panti and Mainamaji are exposed to excess nickel in their drinking water. This can cause decreased body weight, heart and liver damage, and skin irritation.

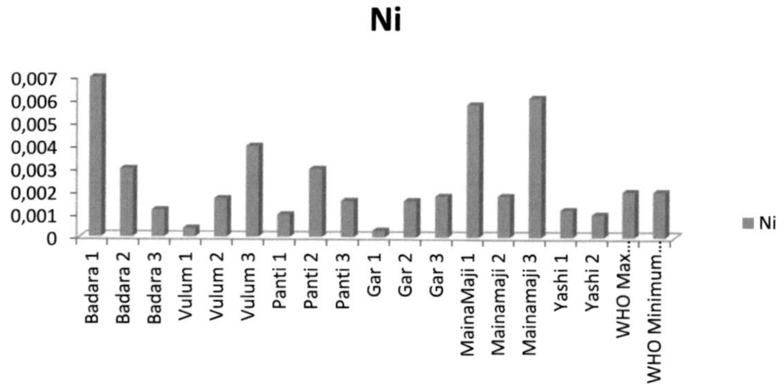

Figure 11. Chart showing the concentration of Nickel in drinking water of the study area.

However people living around Gar and Yashi villages have nickel being deficient in their drinking water.

Conclusion.

Trace elements are chemical components that naturally occur in groundwater plant, and wildlife in minute concentrations. Also known as trace minerals, trace elements are necessary for the optimal development and metabolic functioning of all living things; however at higher or lower concentrations they can cause serious health implications. The concentrations of chromium and nickel in some locations of the study area were found to be above the maximum permissible limits set by WHO. The concentration of lead, copper, manganese, zinc and calcium were found to be below the minimum permissible limit set by WHO.

References.

1. Bata T. P, Dike E.F.C, Samaila N.K, and Bassi D.A (2009). Paleocurrent, Petrography and Provenance analyses of the Kerri-Kerri Formation along Mainamaji-Duguri Road Bauchi state Upper Benue Trough. League of Researchers in Nigeria Vol. 10, No.1 pp 44 - 48

2. Benkhelil. M. J. (1989). The Origin and Evolution of the Cretaceous Benue Trough. Nigerian Journal of African Earth Science 8 . Pp 151 – 228.

3. Carter JD, Barber W, Taitt EA, Jones GP (1963) The Geology of Parts of Adamawa, Bauchi and Borno provinces in Northeastern. Geol Surv. Nigeria Bull 30: 108pp

4. Edmunds, W.M (1996): Geological framework for water quality studies in Sub-Sahara Africa. Journal of Africa Earth Science vol. 22 (4) pp. 385 – 389.

5. Free Drinking Water (2010):http://www.freedrinkingwater.com/water quality/chemical/water -problems-manganese.htm

6. Lenntech Water Treatment Solutions, (2009): http://www.lenntech.com/processes/heavy/heavy-metals/heavy-metals.htm#ixzz12LBu54sg

7. Obaje, N.G. (2009); Geology and Mineral Resources of Nigeria, lecture notes in earth sciences. Springer-Verlag 218p.

8. Skipton, S.; Hay, D.(1988) Drinking Water: Copper. Nebraska Cooperative Extension Publication G 98 – 1360-A Licoln, Nebraska.

9. Skoog, D.A., West D.M.and Holler F.J.(1992): Fundamentals of Analytical Chemistry: Saunders collage publishing, fort worth , US 1992.

10. WHO (1997). Guidelines for Drinking Water quality. Health Criteria and Other Supporting information. Vol.1, 2nd Ed. World Health Organization, Geneva.